OUR AMAZING PLANET

INTRODUCTION TO GEOGRAPHY
for Young Scientists

OUR AMAZING PLANET

INTRODUCTION TO GEOGRAPHY
for Young Scientists

Published by
Heron Books, Inc.
20950 SW Rock Creek Road
Sheridan, OR 97378

heronbooks.com

Special thanks to all the teachers and students who
provided feedback instrumental to this edition.

Fifth Edition © 1993, 2022 Heron Books.
All Rights Reserved

ISBN: 978-0-89-739286-0

Any unauthorized copying, translation, duplication or distribution, in whole or in part, by any means, including electronic copying, storage or transmission, is a violation of applicable laws.

The Heron Books name and the heron bird symbol are registered trademarks of Delphi Schools, Inc.

Printed in the USA

18 July 2022

At Heron Books, we think learning should be engaging and fun. It should be hands-on and allow students to move at their own pace.

To facilitate this we have created a learning guide that will help any student progress through this book, chapter by chapter, with confidence and interest.

Get learning guides at
heronbooks.com/learningguides.

For teacher resources,
such as a final exam, email
teacherresources@heronbooks.com.

We would love to hear from you!
Email us at *feedback@heronbooks.com.*

Your YOUNG SCIENTIST JOURNAL

Scientists love to explore the world and how things in it work. They like to go new places and discover things they've never seen before.

They also like to keep track of what they find. They often fill books with notes and drawings of what they see, and include their thoughts and questions about it. These books are called *science journals*.

What's fun about a science journal is that you can use it to draw pictures or sketches of things that interest you. You can write down ideas you have about things, make maps, write down questions you have and things you want to find out more about. You might even stick in it samples of things you find—flowers, bugs, leaves, feathers, spider's webs—who knows what?

The learning guide that goes with this book will sometimes ask you to look at things and make notes or drawings in a journal of your own.

Whatever you put in your science journal, it will be full of your own personal discoveries. No two journals are alike.

You can use a journal like the one shown here, or you can use a notebook of your choice. You might even want to make your own science journal and use that.

Whichever type of journal you choose, it will be a place to keep drawings and notes about what you are finding out about the world and how it works.

So get ahold of a science journal, or make one, and then get going to see what you can find out. Who knows what might be waiting for you?

IN THIS BOOK

1 STUDYING THE EARTH — 1

2 THE EARTH'S CRUST — 6

3 WHAT SHAPES THE EARTH? — 10
  Plates — 12
  Weathering — 15
  Erosion — 17
  Humans — 19
  Let's Do This: Take a Look Around — 21

4 THE LAND — 22
  Mountains — 22
  Hills — 24
  Volcanoes — 24
  Valleys — 25
  Plains — 26
  Plateaus — 26
  Canyons — 28
  Caves — 29
  Island — 30
  Peninsula — 31
  Let's Do This: Make a Continent Part 1 — 33

5 THE WATER — 34
  Oceans — 34
  Seas — 36
  Lakes and Ponds — 37
  Rivers and Streams — 38

	Wetlands	39
	Deltas	40
	Glaciers	42
	How Land And Water Features Fit Together	44
	Let's Do This: Make a Continent Part 2	47

6 WHAT MAKES THE SEASONS? 48
- The Earth Circles the Sun — 48
- The Earth Spins at the Same Time — 49
- The Earth's Axis — 50
- The Earth's Axis Is Tilted — 50
- The Earth's Tilt Makes the Seasons — 51
- Let's Do This: What Makes the Seasons? — 54

7 CLIMATE 56
- The Hottest Climate Zone — 58
- The Coldest Climate Zone — 60
- Seasons in the Polar Zones — 64
- Between the Hot and Cold Zones — 66
- Let's Do This: Climate Zones — 70

8 LIVING THINGS 72
- Tundra — 76
- Forest — 78
- Taiga — 80
- Temperate Forest — 82
- Tropical Rainforest — 86
- Chaparral — 89
- Grassland — 90
- Desert — 92

9 OUR AMAZING PLANET 94
- Let's Do This: Where Is It? — 96
- Let's Do This: Explore a Place — 98

MAP– EARTH'S BIOMES   100

CHAPTER 1
STUDYING THE EARTH

Humans have lived on Earth for a long time. And all that time they've been learning more and more about it.

Many thousands of years ago people had to learn where to find shelter and safety, where they could find animals to hunt, which plants were safe to eat, and which weren't.

Gradually they began to find out more about the earth. They learned which parts were mountainous, where they could find rivers and lakes, and which parts were deserts.

Later they learned to make maps of the continents and oceans.

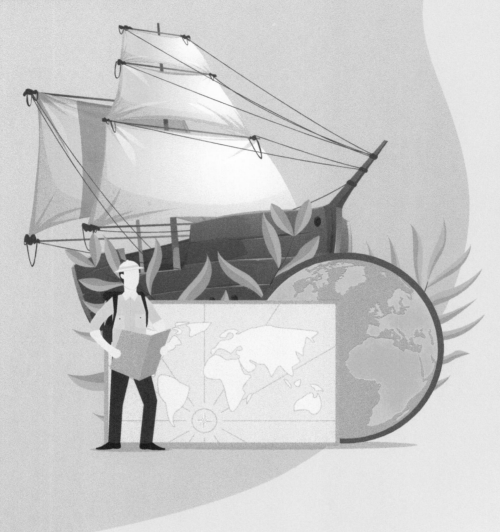

As they traveled, they came to understand that the earth was round, not flat.

They learned what caused the seasons, and what the weather was like in many different parts of the world.

All this knowledge is part of a science we call **geography**.

When we learn about the earth, where its continents, oceans, rivers, mountains, deserts and countries are, we are studying geography.

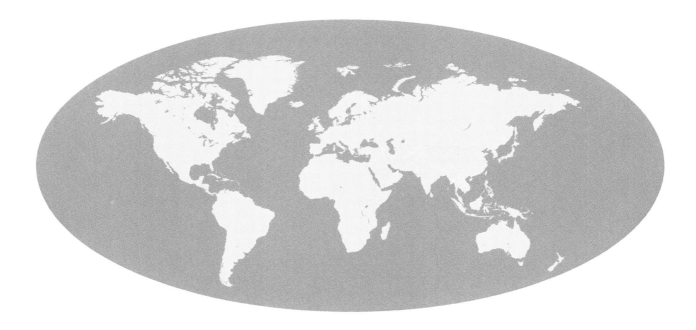

We also learn where people live on the earth and what they do there. Do they live in forests, or deserts? Do they live in houses, tents, or igloos? Do they grow food? Do they hunt or fish? Do they drive sleds or ride camels?

Some live where it almost never rains.
How do they find enough water?

Others live where it rains almost
all the time. How do they protect
themselves from floods?

Some live in places that are extremely hot. How do they build houses that will keep them cool?

Some live in places that are extremely cold. How do they find food when the ground is covered with snow?

Scientists who study the earth and its people, are called **geographers**.

The earth is our home. As a young scientist, there is a lot you can learn about the planet we all live on. You, too, can be a geographer!

CHAPTER 2: THE EARTH'S CRUST

The big round ball we call Earth is made up of layers. Even though we're walking around on them all the time, we only see the top one.

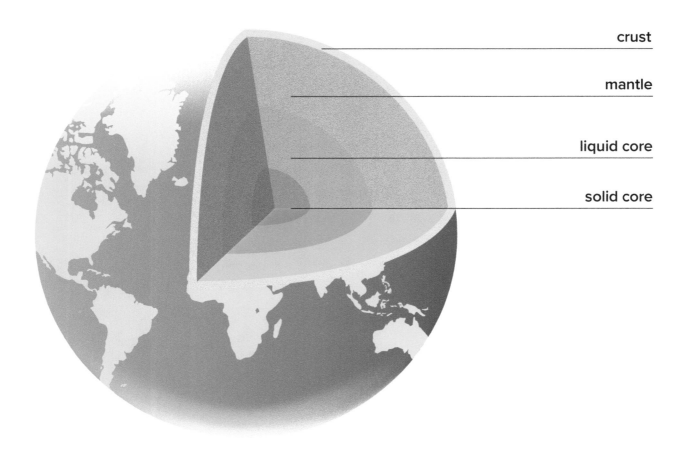

- crust
- mantle
- liquid core
- solid core

At its very center is what we call the earth's **core**. This is a huge ball of solid metal, mostly iron, surrounded by a layer of liquid metal. The core is incredibly hot. Its temperature is 9800 degrees, which is as hot as the surface of the sun!

Around the core is another layer we call the **mantle**. This is a layer of rock about 1800 miles thick. That's about the distance from New York City to Denver.

The mantle is cooler than the core. It has several layers. Each layer is cooler than the one beneath it. The layer next to the core is hot melted rock. The layer at the top is mostly solid rock.

The very top layer of our planet is the thinnest. It is a layer of rock we call the **crust**. The earth's crust is like the skin of an orange, a thin top layer surrounding what's underneath.

Let's take a closer look at the earth's crust. This is the layer we see when we look around. It is where we live, along with all the earth's plants and animals.

In some places the crust is flat for miles and miles, while in other places it is full of high mountains.

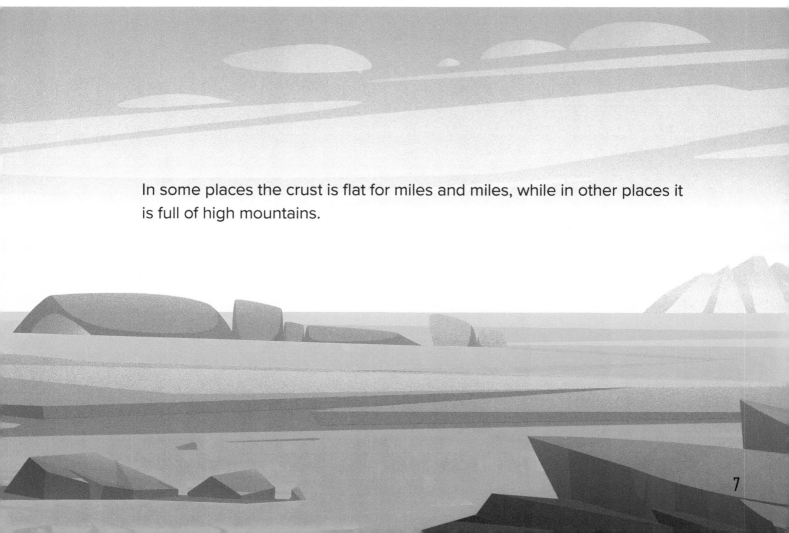

It has low places like the Grand Canyon, dry places like Egypt in Africa, and wet places like the Everglades in Florida. All of these, and more, are what we call features of Earth.

A **feature** is part of something that stands out because it's different.

A hill is different from the valley it's next to. It stands out.

A river, for example, is different from the field it runs across, so it stands out.

When you go outside and look around, you'll see that the crust of the earth we live on has many different features. You will find hills and valleys, lakes, rivers, oceans, islands and more.

Let's see what we can learn about some of these.

CHAPTER 3: WHAT SHAPES THE EARTH?

Most of the earth's features we see around us seem like they've been there forever. Or at least for a long, long time. And many of them have been.

But the fact is, the earth's crust never stays the same. It's always changing. The wind moves soil around. Rains cause flooding and a new stream is born. The earth's features have been changing ever since the crust was first formed over 3 billion years ago. That's a long time and a lot of change!

Some of the changes happen very, very slowly. For example, rain and wind blow over a steep, rocky mountain range again and again for millions of years. Ever so slowly, the rock is worn down and the mountains become smaller.

Some of the changes to the earth's surface are quick. One day a volcano erupts and lava pours out over the land.

Another day, somewhere else, there is an earthquake, the land shakes and a split opens in the earth.

Let's take a look at some things that change the surface of the earth every day.

PLATES

Did you know that millions and millions of years ago all the continents of the earth were joined together in one big piece of land? Scientists named this big piece of land **Pangea** (pan JEE uh).

About 175 million years ago, Pangea began to break apart into several large pieces. You can see this when you look at the east coast of South America and the west coast of Africa. They almost fit together like pieces of a puzzle.

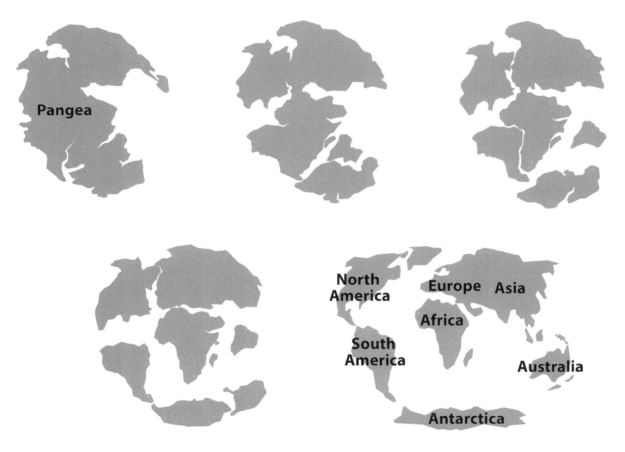

Our continents are still moving ever so slowly apart, perhaps an inch or two a year. Why is this?

The outer layer of the earth, the crust plus the top layer of the mantle, is broken into several gigantic slabs of rock called **plates**. There are seven large plates, along with some smaller ones. They all fit together like pieces of a big puzzle that wraps around the earth.

The Main Plates

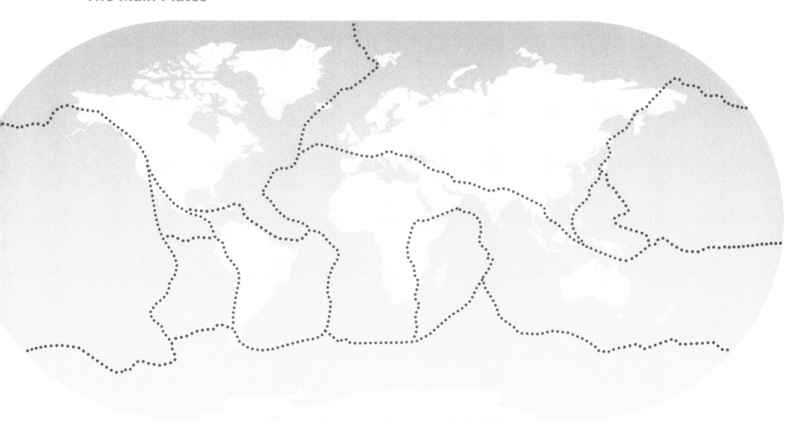

The mantle layer that lies just under the crust is partly solid and partly liquid rock. Because it's partly liquid, it moves and slides around some. So the plates are always moving a little bit too. They move towards or away from one another, tiny bit by tiny bit.

In science these are called **tectonic** (tek TAHN ik) **plates**. "Tectonic" has to do with changes to the surface of the earth. When tectonic plates move, they cause changes in the earth's crust.

When two plates collide, it can push part of the earth's crust upward to create hills or mountains.

Fifty-five million years ago, the tallest mountains in the world, the Himalayas, were created when two tectonic plates slowly collided and pushed huge chunks of the earth's crust upwards.

The plates have continued to push against each other ever since, slowly raising the mountains higher and higher.

Occasionally the edges of two plates rub against one another. When this happens we might experience an earthquake.

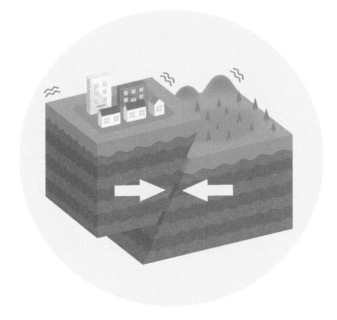

Earthquakes can cause landslides, where large amounts of rock and dirt fall or slide downward.

Tectonic plates are always moving, very, very slowly. They move so slowly that we usually can't feel the motion. But it goes on all the time.

WEATHERING

Weather shapes the surface of the earth. It's been doing this for millions and millions of years. This is called **weathering**.

Rain, wind and ice slowly dissolve rocks, or break them down into smaller pieces. This takes a long time. Rain, for example, can cause some of the minerals in a rock to slowly dissolve, and gradually the rock breaks apart into many smaller pieces.

Wind can blow small grains of sand against a large rock and very gradually wear it away. Rocks have also been worn away by waves until they are round and smooth.

Sometimes water gets into a crack in a rock, then freezes when the weather gets cold.

Ice takes up more space than water. So the ice widens the crack.

After many years of this, the rock gradually breaks apart into smaller pieces.

The earth's crust was originally a more solid layer of rock. Rock is still its main ingredient. For millions of years, weathering has been breaking all this rock into smaller and smaller pieces. It's still happening, and will continue to happen for a long, long time.

EROSION

Here is another way the surface of the earth is changed. Wind, water and ice will often pick up rock and soil and carry it to different places. This is called **erosion** (i ROH zhun).

The material that is picked up and moved is called **sediment** (SED uh munt). Sediment can contain large rocks and tiny grains of sand. Gravel, pebbles and soil are often part of sediment.

Water streaming down a hillside after a heavy rain will pick up sediment along the way and carry it somewhere, perhaps to a creek or river. This is erosion.

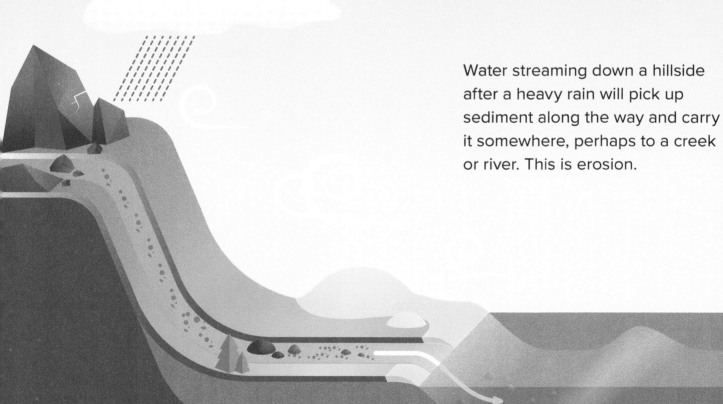

Bit by bit, waves can carry away the sand on a beach. This gradually makes the beach smaller and smaller. That's erosion.

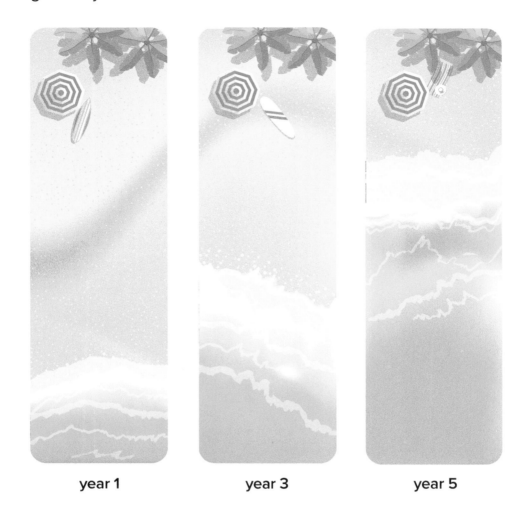

year 1   year 3   year 5

Wind can pick up soil from a field and carry it away. The field now has less soil and some other place has more.

The bigger and stronger the wind or water flow, the more erosion there is.

All of these things, the motion of the earth's tectonic plates, weathering and erosion, shape the earth's features you see around you every day.

HUMANS

Humans have been making changes to the earth's surface for a long time. The American pioneers, for example, cut down forests and turned huge amounts of land into fields for farming.

Using science, man has been able to change the earth's surface in many ways. For example, many countries around the world build dams that create huge lakes.

Today in some parts of the world, forests are being cut down to make room for homes and livestock.

In other parts of the world, plains are being turned back into forests by planting thousands of trees.

These are some examples. There are more. Humans are making changes to the earth's surface all the time.

TAKE A LOOK AROUND

For this activity you will need

- your science journal
- pencil

Steps

1. Explore outside. You might walk around your neighborhood, take a nature hike, or explore a park. See if you can find signs of changes to the earth's surface caused by humans.

2. Make notes in your science journal about what you find. Include what you found, where you found it and what it looked like. Make some sketches of what you see.

3. Find as many examples of changes to the earth's surface as you can.

CHAPTER 4: THE LAND

The earth's features take many different shapes. Some, like mountains, are found on land. Some, like lakes, are water features. But we call them all **landforms**.

In this chapter, we will talk about some of the largest land features. You can find most of these on every continent.

MOUNTAINS

A landform that rises about a thousand feet or more above the land around it is called a **mountain**. A group of mountains strung together in a line is called a **mountain range**. For example, the Himalaya mountains form a mountain range.

Mountains take millions of years to form. In many cases they are formed by the slow collision of tectonic plates. The Rocky Mountains in North America and the Andes Mountains in South America were formed this way.

Mountains that are sharp with steep sides are younger. Older mountains are more rounded because they've been worn down by weathering.

In many parts of the world you will find mountains that are more than 10,000 feet high. That's almost two miles. The world's highest mountain, Mount Everest, is over 29,000 feet high. That's over 5 miles!

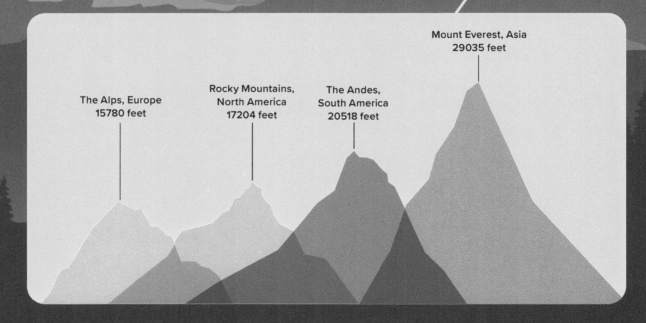

The Alps, Europe
15780 feet

Rocky Mountains, North America
17204 feet

The Andes, South America
20518 feet

Mount Everest, Asia
29035 feet

HILLS

A **hill** rises above the land around it. It has sloping sides. Hills are not as high or steep as mountains.

VOLCANOES

A volcano happens when hot liquid rock from the earth's mantle rises up and bursts out through a break in the earth's crust. When this happens we say the volcano is **erupting**.

Volcanoes mostly happen where the edges of two tectonic plates meet one another. Most of the earth's volcanoes erupt thousands of feet beneath the surface of the ocean.

VALLEYS

A long area of low land that lies between hills or mountains is called a **valley**. Often a valley has a river or stream running through it.

The hot liquid rock is called **magma**. When it pours out, we call it **lava**.

The lava hardens when it cools, and gives the volcano its shape.

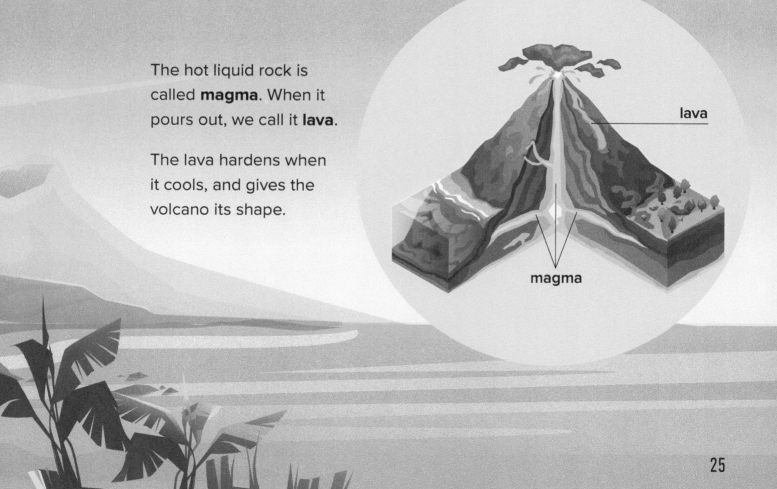

PLAINS

Plains are large areas of flat land with very few trees. There may be a few rolling hills here and there.

More than 1/3 of the land on the earth's crust is plains. There are plains in the far north, plains in Africa, plains in Russia, plains in the United States and Australia. Many are covered with grass.

PLATEAUS

A **plateau** (pla TOH) is a large, flat area of land that rises up higher than the land around it. It usually goes up steeply on at least one side.

There are some enormous plateaus in Asia, Africa, North America, and even Antarctica. The western part of the United States also has a number of smaller plateaus.

CANYONS

A **canyon** is a deep narrow valley with steep sides. Often a river runs along the bottom.

The Grand Canyon in Arizona is an enormous canyon over 250 miles long. In some places it is a mile deep. Several million years ago, the Colorado River ran across the top of a huge plateau. Gradually the river washed away more and more of the rock. Now, after millions of years of erosion, we have a long, deep canyon with the Colorado River running along the bottom.

CAVES

A cave is a hollow space under the ground or inside another landform, such as a mountain. A cave can be large enough for a person to enter.

Some caves are created when water dissolves the minerals in certain kinds of soft rocks. Over a long, long period of time the rocks gradually disappear, leaving a cave, or even a whole group of caves.

Caves can also be created by wind and water wearing away at rocks along a seashore.

ISLAND

An **island** is a piece of land surrounded by water. Islands are found in the ocean, in lakes, and rivers.

There are a huge number of islands on the earth, almost too many to count.

Some, like Cuba and Iceland, are quite large. Others are tiny.

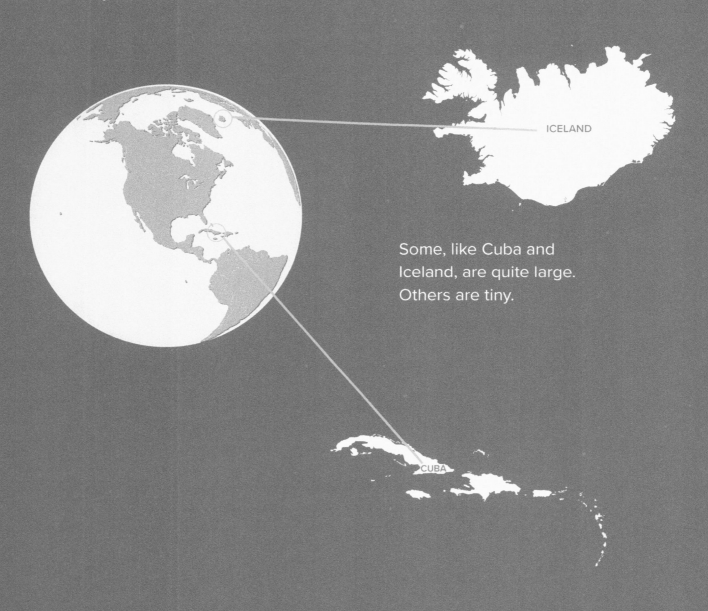

PENINSULA

A **peninsula** is a piece of land that sticks out from a larger area of land and is almost surrounded by water, but not quite. A peninsula has water on all sides except one, and this side is connected to a larger area of land.

Some peninsulas are small, sticking out less than a mile into the water.

Some, like Cape Cod in Massachusetts, are larger. Cape Cod reaches out into the ocean over 50 miles.

Other peninsulas are very large. Most of the state of Florida is a huge peninsula, nearly 500 miles long.

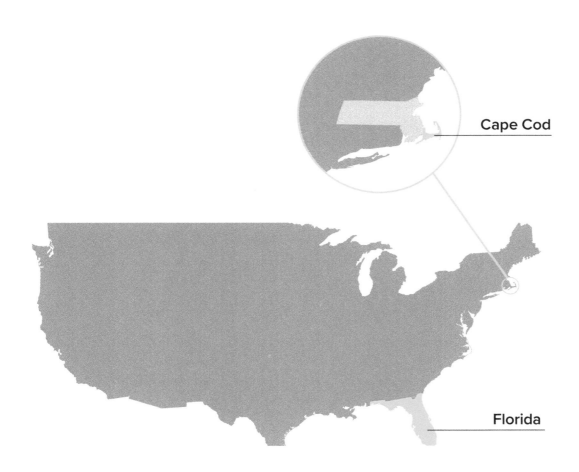

MAKE A CONTINENT PART 1

For this activity you will need:

- modeling clay, several colors
- large, flat surface that won't be harmed by the clay

Steps

1. Make a large model of a continent that shows these things:
   - a plain
   - one or more plateaus
   - mountains
   - hills
   - one or more valleys
   - any other landforms you would like to include

2. Label each landform you make. If you want to, you can make up names for them like "The Blue Hills" or "Oak Valley."

3. Give your continent a name. Save your model for a later activity.

CHAPTER 5: THE WATER

Three-fourths of the earth's surface is covered with water. There's so much water that Earth is sometimes called "the blue planet." From space it looks mostly blue. Let's find out where all this water is by reviewing Earth's major water features.

OCEANS

Pick up a globe and look at it. Turn it around and around. You will see a huge area of saltwater surrounding all the continents. This is called the **ocean**, or the **sea**. It holds most of the earth's water.

This huge ocean divides into five smaller oceans, each with its own name.

ATLANTIC OCEAN

PACIFIC OCEAN

The largest of these is the **Pacific Ocean**. It contains almost half the water on Earth.

The next largest, and the saltiest, is the **Atlantic Ocean**. Beneath its surface lies a huge underwater mountain range that stretches all the way from Iceland to the tip of South America. This mountain range has formed where two tectonic plates are pulling apart.

ARCTIC OCEAN

The **Arctic Ocean** is the smallest and coldest of the oceans. It's so cold that it has almost no plants. It is full of fish, whales and other animals. For much of the year, the Arctic Ocean is covered with a thick layer of ice.

PACIFIC OCEAN

INDIAN OCEAN

The **Indian Ocean** is the third largest, and also the youngest of the earth's oceans. It was formed when India, Africa and Australia slowly split apart around 50 million years ago.

SOUTHERN OCEAN

The **Southern Ocean** surrounds Antarctica. It is home to animals such as whales, penguins and seals.

We don't know a lot about the Southern Ocean because it hasn't been explored as much as the others. More and more new things about it are now being discovered.

There are a lot of things we don't know about all our oceans. For example, very little of the ocean bottom, called its **floor**, has been mapped. There's still a lot for scientists to explore and find out.

SEAS

A sea is an area of saltwater smaller than an ocean and mostly surrounded by land. The Mediterranean Sea (med uh ter AYN ee un) between Europe and Africa is a good example.

LAKES AND PONDS

A **lake** is a large area of water surrounded by land. Most lakes are full of fresh (not salty) water.

Scientists estimate that there are 117 million lakes on planet Earth! Canada, with 880,000 lakes, has the most of any country in the world.

Some lakes are so big you can't see from one side to the other. For example, Lake Superior in North America is 350 miles from one end to the other. In fact, it takes up more space than the fives states of Rhode Island, Delaware, Connecticut, Hawaii and New Jersey combined!

A very small lake is called a **pond**.

RIVERS AND STREAMS

When rain falls on mountains or hills, it runs down to lower ground. Water from melting snow also runs down. Moving water like this is called a **stream**.

The starting point of a stream is called its **source**.

Small streams often flow into larger streams, and these get wider as they join up. Eventually they become a **river**, a large flow of water.

A river flows along and eventually pours into a lake, a larger river, or the ocean. This part of a river is called its **mouth**.

Rivers provide water to villages, towns and cities. Because they flow along and often go to the ocean, they have always been an important way for people to travel to other places.

WETLANDS

Wetlands are areas of the earth where water covers the ground and doesn't drain away. The water in a wetland can be saltwater, fresh water, or a combination. The land is wet and soggy most of the time.

Wetlands are important because they act as huge water filters that suck pollution out of the world's water. They are also home to many kinds of plants, animals, and birds.

A wetland with lots of trees is called a **swamp**.

A wetland with mainly grass and low bushes is called a **marsh**.

Wetlands are found in many places on the earth, most often near rivers, lakes and oceans. The Everglades in Florida is one of the world's largest wetlands.

DELTAS

Have you ever seen a river full of brown, muddy-looking water? A river, especially when it's full and fast moving, can pick up a lot of soil and rocks as it moves along. These make it muddy.

The muddy water flows along and eventually reaches the mouth of the river.

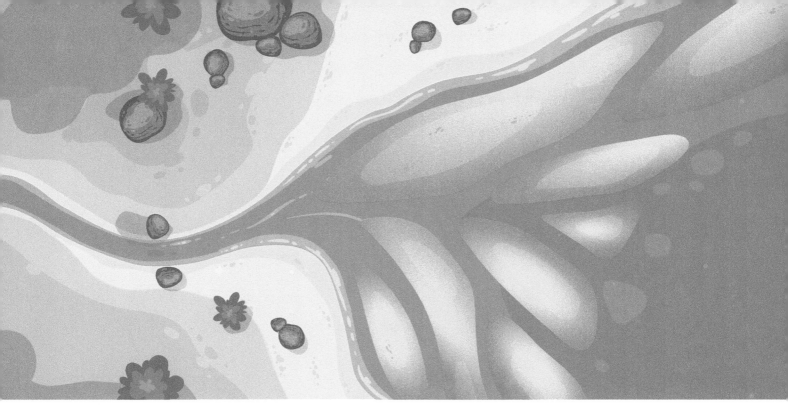

As it pours into the ocean, the water spreads out. And as it does this, it slows down. The soil and rocks in the water slow down, too. They fall to the bottom and are deposited, or left there. Eventually, a fan-shaped area of deposited soil builds up. We call this a **river delta**, or just a **delta**.

After many years, the river begins to cut smaller streams through the built-up soil of the delta.

The delta at the mouth of the Mississippi River is over 60 miles wide. It took thousands of years for it to grow to this size.

The soil in a delta is often great for growing food. In Egypt, for example, the rich soil of the Nile River delta has been a center of farming for over 5000 years.

GLACIERS

Glaciers are enormous masses of snow and ice that build up in places where it is so cold they can't melt away. Over time, these piles of ice and snow get deeper and deeper. Some glaciers are over a mile deep!

As layers of ice and snow build up, they become very heavy. This makes the glacier begin to flow to lower ground. Glaciers are like frozen rivers, flowing for miles.

Glaciers move very slowly. Some move 100 feet in a day, but most move only a foot or so. A small glacier may move only a foot or two per year!

A glacier is strong enough to pick up huge boulders and piles of rock and soil. As it flows along, it scrapes away the earth beneath it and leaves the land flat. It crawls along, grinding and crushing things in its path. Many of the world's lakes were created by glaciers. A glacier can scoop out a hollow space that later fills up with water, creating a lake.

When a glacier reaches a lake or ocean, gigantic chunks of ice can break off and crash into the water. This is how icebergs are born.

Thousands of years ago, a third of the earth was covered by glaciers. When these finally melted, they had formed and left behind many of the mountains, valleys, lakes and plains that we see around us today.

HOW LAND AND WATER FEATURES FIT TOGETHER

Here are most of the land and water features we've talked about so far, showing how they might fit together.

mountains

plateau

canyon

hill

lake

island

delta

MAKE A CONTINENT PART 2

For this activity you will need:

- the continent model you made in the Make a Continent Part 1 activity
- modeling clay, several colors

Steps

1. Add these water features to your continent:
   - one or more oceans
   - one or more lakes
   - one or more rivers
   - a delta at the mouth of a river
   - any other water features you learned about that you would like to add

2. Label each feature you make. If you want to, you can make up names for them like "Pebble River" or "Lake Lisa."

3. Show your continent to another person and answer any questions they have about it.

CHAPTER 6: WHAT MAKES THE SEASONS?

As you may know, the seasons don't happen at the same time everywhere on Earth. When it's winter in North America, it's summer in South America. When it's winter in South America, it's summer in North America. Why are the seasons different from place to place?

To understand how seasons happen, let's start by looking at the earth from space.

THE EARTH CIRCLES THE SUN

From space we would see that the earth is always circling around the sun. We say it **revolves** around the sun, always moving around it in a big circle, called an **orbit**.

The amount of time it takes for the earth to circle the sun once is what we call a year.

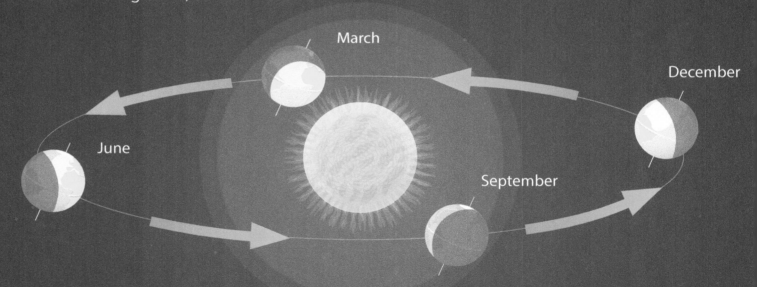

This drawing and the others in this chapter make the earth look much larger and closer to the sun than it really is. The earth is 93 million miles away from the sun. To show this distance realistically, this page would have to be over 1000 feet wide. And the sun would have to be about five feet tall!

THE EARTH SPINS AT THE SAME TIME

The whole time the earth is making its year-long trip around the sun, it is *also* **rotating**, meaning it is spinning around and around like a top, making a full spin every 24 hours.

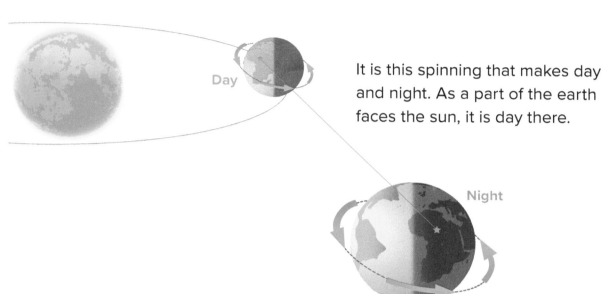

It is this spinning that makes day and night. As a part of the earth faces the sun, it is day there.

As the earth continues to spin, that same part of the earth turns away from the sun and it becomes night there.

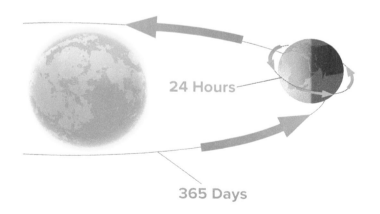

Then, as the earth continues to spin, it becomes day again. That single spin took 24 hours.

So there are two things happening at the same time. Every day the earth rotates, making a full spin. As it rotates, it also revolves around the sun in a big circle. This takes 365 days, or a year.

THE EARTH'S AXIS

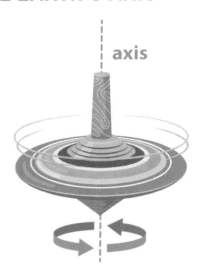

An imaginary line through the center of a spinning object, around which it seems to turn, is called its **axis** (AK sis).

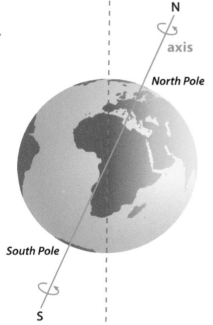

Since it is a spinning object, the earth has an axis. This is the imaginary line that runs through the earth's center, from the North Pole to the South Pole. The earth rotates around this axis.

THE EARTH'S AXIS IS TILTED

Looking at the picture above you will see that the earth's axis doesn't go straight up and down, it is tilted. As the earth circles the sun, the tilt of its axis doesn't change. It stays the same.

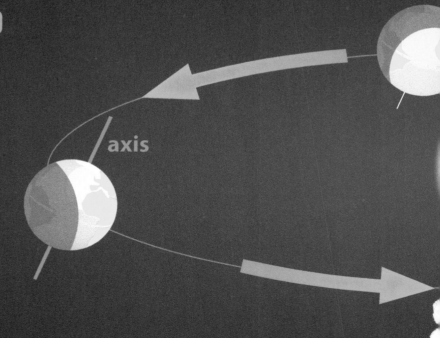

50

THE EARTH'S TILT MAKES THE SEASONS

It's the tilt of the earth that gives us our seasons. Here's how this works.

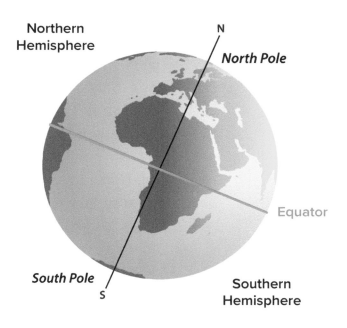

There is an imaginary line circling the middle of the earth. We call this the **equator**. Because it's imaginary, you won't see it on the earth itself, but you will see it on most globes and world maps.

The equator divides the earth into two equal parts called **hemispheres**, which means "half a sphere." These two parts are the Northern Hemisphere and the Southern Hemisphere.

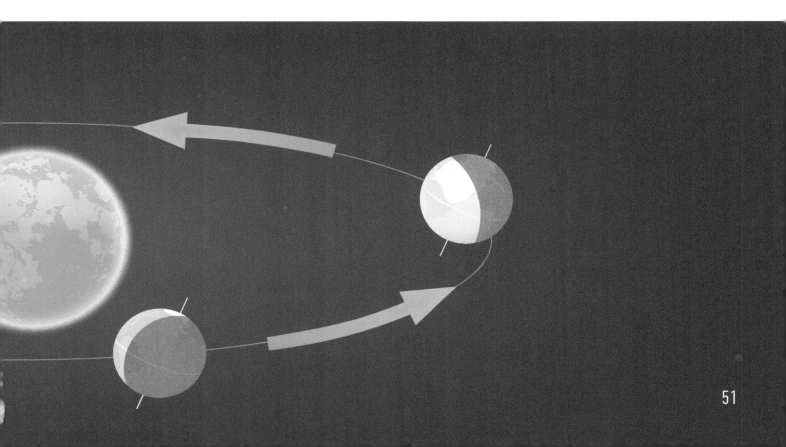

Because the earth is tilted, the two hemispheres don't get the same amount of sunlight. The hemisphere tilted toward the sun gets more direct sunlight and is warmer. The hemisphere tilted away from the sun gets less sunlight and is colder.

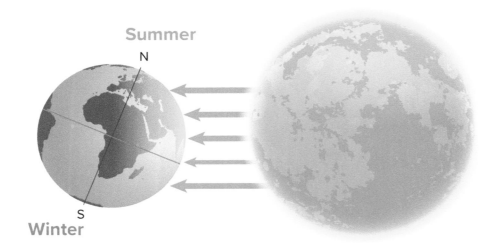

During part of the year, the Northern Hemisphere is tilted *toward* the sun. At this time the Northern Hemisphere is warm and sunny, so it's summer.

But because its tilt does not change, by the time the earth has gotten halfway around the sun, the Northern Hemisphere is tilted *away* from the sun. This makes the Northern Hemisphere colder. It's winter.

The Southern Hemisphere is tilted *toward* the sun. This makes it warmer. It's summer.

This means the seasons in the two hemispheres are opposite from one another. One is cooler when the other is warmer.

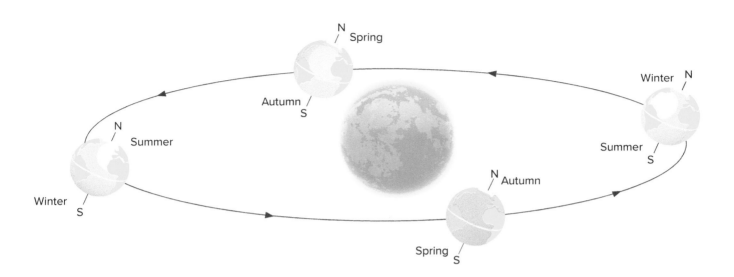

In the Northern Hemisphere, summer is in June, July and August. But during these months, the Southern Hemisphere is experiencing winter. In the Northern Hemisphere winter is in December, January and February. During these months the Southern Hemisphere is experiencing summer.

As the earth travels around the sun year after year, the seasons in both hemispheres repeat again and again. And in the Northern Hemisphere they're always the opposite of what they are in the Southern Hemisphere.

So, if you live in the United States (Northern Hemisphere), when do you want to travel to Australia (Southern Hemisphere)? Well, that depends on if you want to go surfing or snowboarding!

And if you live in Australia and like to surf, when winter comes, just head north to warmer weather in France or California, two popular places for surfing!

WHAT MAKES THE SEASONS?

For this activity you will need

- globe
- small lamp without a shade
- partner to work with

Steps

1. Put a small lamp without a shade in the center of a room. This will stand for the sun. Place a globe several feet away. Make sure the light bulb and the center of the globe are at the same height.

2. Decide which way you want the globe's axis to point.

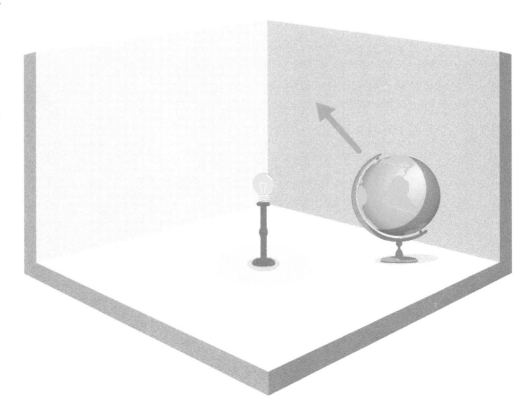

③ Turn the lamp on. Turn the other lights in the room off.

④ Move the globe (earth) around the sun, keeping its axis tilted toward the corner you chose. As you do this, stop every once in a while to notice which hemisphere is getting the strongest "sunlight" from the lamp.

⑤ Find the place in the earth's trip around the sun where the North Pole is closest to the sun. Explain to a partner or teacher why this is summer for the Northern Hemisphere and winter for the Southern Hemisphere.

⑥ Find the place in the earth's trip around the sun where the North Pole is farthest from the sun. Explain to a partner or teacher why this is winter for the Northern Hemisphere and summer for the Southern Hemisphere.

⑦ Move the globe around the sun again so that it is in the December position for the Northern Hemisphere.

⑧ On the globe, find a place in the Northern Hemisphere. Think of an activity people there might be doing in December. Now find a place in the Southern Hemisphere. Think of an activity people might be doing there in December. Explain to a partner or your teacher what you thought of.

CHAPTER 7: CLIMATE

People tend to pay attention to the weather. Will tomorrow be sunny and warm so we can go camping? Will my garden get some rain? Will there be enough snow for skiing? Weather can make a big difference in our lives and what we do each day.

Weather is what happens in the sky from day to day. You might have sun on Monday but rain on Tuesday. On Thursday you might have a big snowstorm. **Weather** is what happens in the sky over a short period of time, like day to day, or week to week.

Climate is what the pattern of weather is like over a longer period of time.

For example, is it hot all the time where you live? If so, that is the climate you live in.

Or is it cold and snowy for months and months at a time? That, too, would be the climate you live in.

Where you live, are the four seasons very different from one another? Or is the weather mostly the same a lot of the time? Perhaps where you live, there seems to only be two seasons—a rainy one and a dry one. These are all signs of what the climate of an area is like.

The world has three main climates—a hot one, a cold one, and one that's in between. Each of these climates covers a large section of the earth.

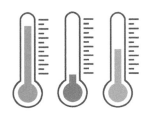

Each part of the earth with the same climate is called a **climate zone**.

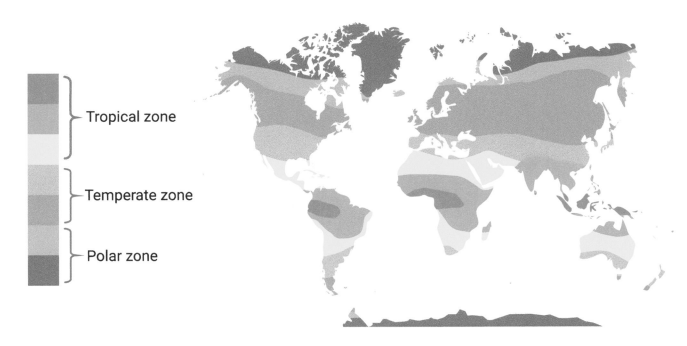

Tropical zone

Temperate zone

Polar zone

57

THE HOTTEST CLIMATE ZONE

Sunlight is one of the biggest things that makes a climate zone the way it is. Does that part of the earth get a lot of sunlight? Does the sun shine directly down on that part of the world?

The **tropical zone** is the part of the earth closest to the equator where the sun shines brightly all year long. This makes it warm or hot there every day. This zone is also called **the tropics**. It is the warmest climate zone.

The tropical zone is quite large. It has a northern boundary and a southern boundary. These are two imaginary lines. The sun shines directly down on the earth in the area between these.

Each imaginary line is called a **tropic**. The line at the northern edge is the **Tropic of Cancer**. The line at the southern edge is the **Tropic of Capricorn**.

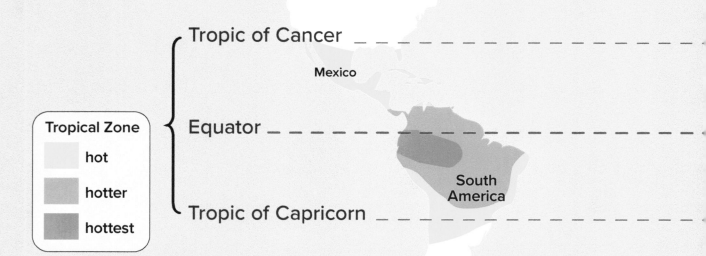

In most areas of the tropics, it rains a lot. There are really only two seasons, a rainy season and a dry season. Both of them are hot. The dry season is usually shorter.

Many of Earth's people live in the tropical zone. If you look at a globe, you'll find many familiar countries in the tropics. Mexico, India, and a number of South American and African countries are there.

Many of the world's animals live in the tropics. You might see, for example, monkeys, parrots, gorillas, jaguars, sloths, toucans, elephants, hippos, rhinos, and many beautiful birds.

Many of our favorite plant foods also grow in the tropical zone. Bananas, mangoes, chocolate, coffee, vanilla, oranges, pineapples and avocados all grow in tropical climates.

THE COLDEST CLIMATE ZONE

Earth's coldest climate is found in what we call the **polar zones**. These are large areas around the North and South Poles.

It is always cold in the polar zones. Of all the places on Earth, these are furthest from the equator. They get sunlight, but it is not strong and direct like it is in the tropics. It's weak and gives little warmth.

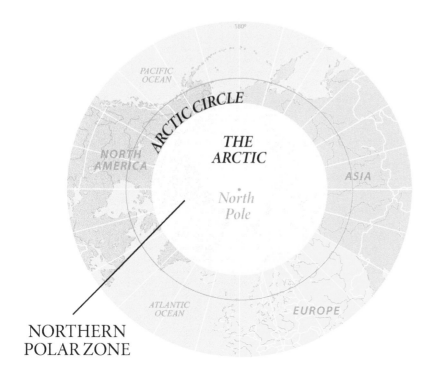

Each of the two polar zones has a boundary.

The boundary of the northern polar zone is called the **Arctic** (ARK tik) **Circle**. The area inside this circle, which includes the North Pole, is called **the Arctic**.

What's in the Arctic? Surrounding the North Pole is a huge sheet of ice many, many feet thick that we call the **polar ice cap**. This floats on a large sea called the Arctic Ocean.

Even though it's very cold, humans have lived in parts of the Arctic for thousands and thousands of years.

The boundary of the southern polar zone is called the **Antarctic** (ant ARK tik) **Circle**. The area inside this circle, which includes the South Pole, is called **the Antarctic**. Antarctic means "opposite to the Arctic."

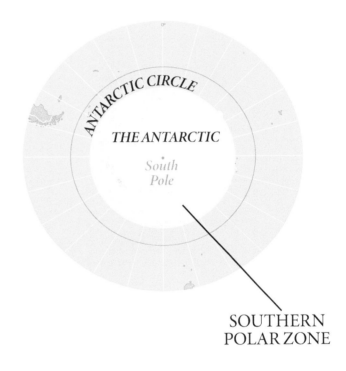

In the Antarctic is a continent we call **Antarctica**. This continent is rock covered with a sheet of ice two miles thick. Surrounding it is the Southern Ocean.

Antarctica is the coldest place on Earth. In winter, the temperature is around 50 degrees below zero! Freezing winds up to 200 miles per hour blow over the ice and snow.

It is so cold that humans have never lived there. Nowadays a few scientists visit Antarctica during the summer to do research.

Being so cold, the polar areas have less animal life than other climate zones. Whales, seals and sea birds live in both the Arctic and the Antarctic. In Antarctica you will also find penguins.

The warmer Arctic has more animal life. There you will find polar bears, walruses, arctic foxes, arctic wolves, snowy owls and reindeer.

SEASONS IN THE POLAR ZONES

The polar zones have only two seasons, summer and winter. Winters are extremely cold and dark. Summers are a bit warmer.

In the Arctic summer, the temperature doesn't often go above 32 degrees. In the Antarctic, it mostly stays below 0 degrees!

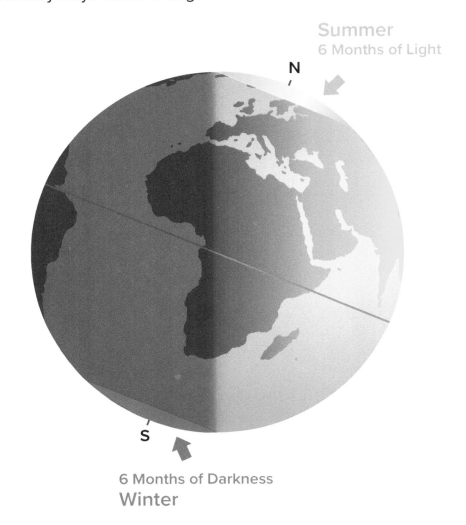

Summer
6 Months of Light

6 Months of Darkness
Winter

An amazing fact is that during summer at the North Pole the sun never sets!

If you look carefully at this picture of the earth showing June in the Northern Hemisphere, you can see why. In June, the North Pole is tilted toward the sun.

So even though the earth is making a full turn, or rotation, every 24 hours, the North Pole is in sunlight all the time. This is true for half the year.

Winter
6 Months of Darkness

6 Months of Light
Summer

In December, this changes and the opposite happens. For many months the North Pole is in darkness. Again, the picture will show you why.

Months of sunlight followed by months of darkness also happens at the South Pole. But it happens at the opposite time of the year from the North Pole.

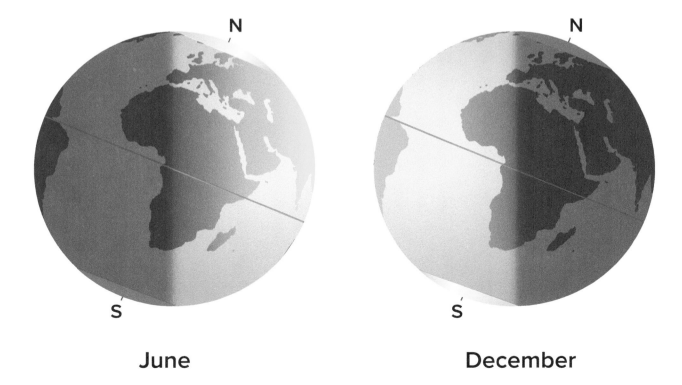

June

December

BETWEEN THE HOT AND COLD ZONES

Between the tropical zone and the polar zones are two zones we call the **temperate zones**. "Temperate" means not extremely hot or extremely cold. That's exactly what a temperate climate is like. It gets warm in summer and cold in winter, but not extremely hot like the tropics or extremely cold like the polar zones.

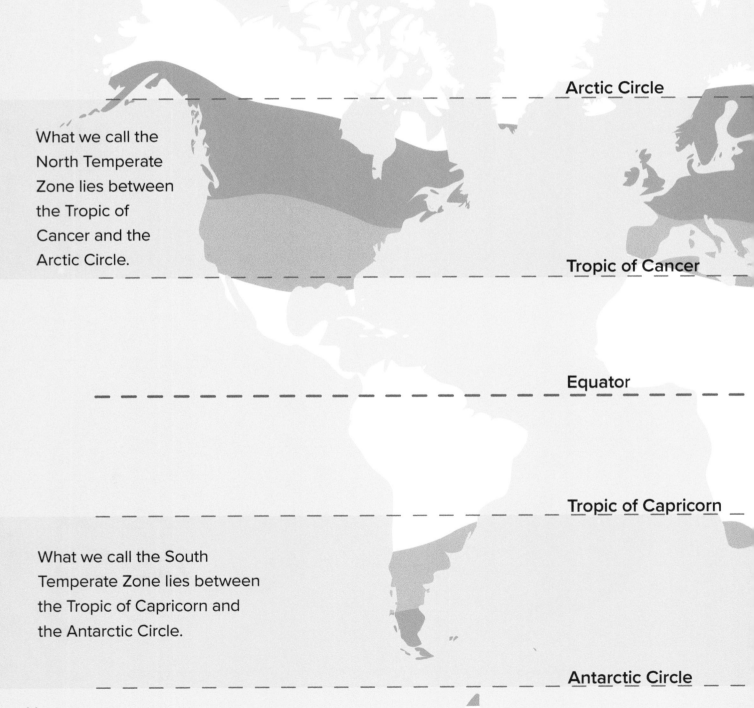

What we call the North Temperate Zone lies between the Tropic of Cancer and the Arctic Circle.

What we call the South Temperate Zone lies between the Tropic of Capricorn and the Antarctic Circle.

These zones, especially the North Temperate Zone, have the most land. Some of the world's biggest countries, the United States, Canada, Russia and China, are all in this zone.

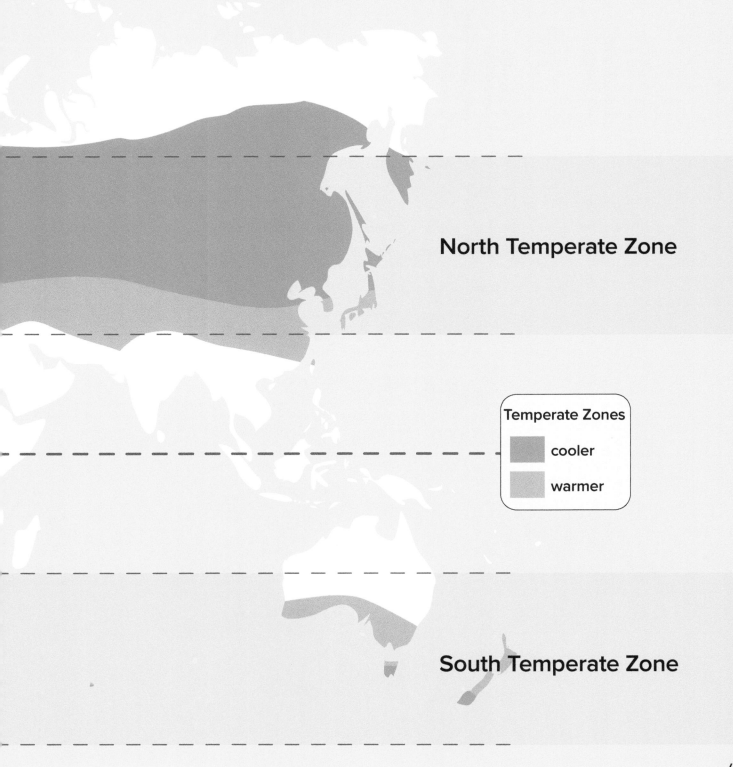

North Temperate Zone

Temperate Zones
- cooler
- warmer

South Temperate Zone

Most of Earth's people live in the two temperate zones.

While the polar and tropical zones have only two seasons, in the temperate zones there are four seasons—summer, fall, winter and spring.

Because they have spring and summer seasons when things can be planted and grown, a lot of farming is done in these parts of the world.

Temperate zones are full of interesting animal life. Bear, otters, buffalo, koala, giant pandas, llamas, foxes, wolves, eagles, owls and many kinds of songbirds are just some of these.

Some of the many foods commonly grown in temperate climates are apples, tomatoes, wheat, nuts, many vegetables like corn, lettuce and broccoli, and other fruits like peaches, pears, plums, strawberries and melons.

CLIMATE ZONES

For this activity you will need:

- globe on a stand
- tape or blank stickers for making labels
- partner to work with

Steps

1. On your globe, find the following and label each one.
   - Tropic of Cancer
   - Tropic of Capricorn
   - Tropical Zone
   - Arctic Circle
   - Arctic
   - Antarctic Circle
   - Antarctic
   - North Temperate Zone
   - South Temperate Zone

2. Have your partner spin the globe. Close your eyes and stop the globe by putting your finger on it. When you open your eyes, say what continent or ocean your finger is on. Then say what climate zone that spot is in.

3. Take turns doing this until it is easy.

4. Now take the labels off and repeat this activity. Continue to take turns until it is easy.

8 CHAPTER LIVING THINGS

So far we've learned a number of interesting things about the planet we live on.

We've gone over what the earth is made of, what its landforms and water features look like, and how its crust is always changing.

We've also learned about different climates and why there are seasons.

But geography gives us something else to think about, and that's what *lives* on the earth. The earth isn't covered with just soil and rocks, but with lots of living things. Many plants, animals and humans all live together on the crust of the earth.

If you were to look down at our planet from space, you would see some large blue areas. These are oceans, rivers, lakes and other areas covered with water.

You would also see many large green areas, green because they are covered with plants.

And you would see some areas that are brown. These are dry areas with few plants.

If you were able to zoom in and look very closely, you would see that certain groups of plants and animals live in each one of these areas.

Let's take a look at why this is.

As you know, each part of the earth has its own climate. The plants and trees that grow there are ones that are suited to how warm or cold it is, and how much rain there is.

A palm tree planted in a hot, dry desert will grow tall and strong.

But plant an apple tree in a desert and it will wither away.

Plant it in a cool climate with sunshine and plenty of rain, and it will grow strong and give you lots of apples.

We also see that the plants growing in an area provide food and shelter for certain animals, but not others.

In a place full of gardens and flowering plants, a hummingbird would find plenty of resting places and all the food it needs.

But in the middle of a dry desert, a hummingbird would have almost nothing to eat and no place to sleep.

A lizard, however, would do well there.

In this way, each area of the earth, with its particular climate, ends up with certain plants and animals that can live there.

There is a name for an area like this. It's called a **biome**. Any large part of the earth with a certain climate and certain plants and animals is a biome. A desert, having a dry climate with certain plants and animals is an example of a biome.

Our planet has several different biomes. Let's take a look at some of the largest ones to see what they are like.

TUNDRA (tun druh)

The first biome we're going to explore is called the **tundra**. This is a vast empty plain in the far north.

The dark purple areas of this map are the tundra.

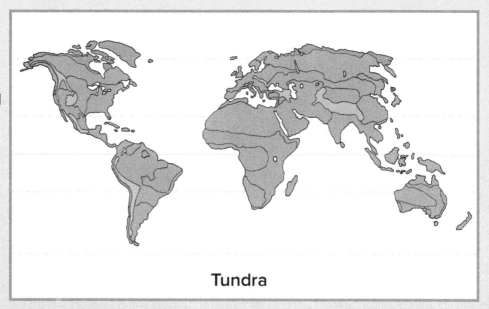
Tundra

Stretching all along the northern edges of North America, Europe and Asia, the tundra is so cold that the ground is almost always frozen.

There are no trees. Tree roots can't grow down into the frozen soil. In fact, tundra means "the place without trees."

The tundra has only two seasons, winter and summer. The winter is long, cold and dark. The temperature is always below zero, and icy winds blow over the frozen land.

The summer is short, only two months, but the sun shines almost 24 hours a day. Plants like mosses, short grasses, wildflowers and small bushes appear, and the tundra comes alive with millions of insects. Many birds migrate here to feed on all the bugs and mosquitoes, and to nest and hatch their young.

When we say "the tundra" we are talking about the huge tundra area up near the North Pole.

There are other smaller areas of the planet that are also tundra. These are areas up near the tops of high mountains all around the earth. The light purple areas of the map are mountain tundra biomes.

Animals living in the tundra must be able to survive long, very cold winters. If you were to fly over the far north, you might see polar bears, musk oxen, white arctic foxes and many caribou, all with thick fur coats.

The edge of Antarctica near the South Pole is also an area of tundra. It's flat and cold with no trees. There you might see penguins, seals and sea lions. These animals have thick layers of fat on their bodies that help keep them warm in the freezing cold temperatures.

FOREST

The earth's biomes don't have exact boundaries the way states and countries do.

A map may make it look like they do, but the change from one biome to the next is gradual.

This means that in some places you may find plants and animals from more than one biome living in the same area.

So, if we were to leave the flat, treeless tundra and travel south, we would gradually begin to pass a few trees, then more and more.

Eventually we would find ourselves in an area covered with trees as far as you can see. This is the plant and animal community we call **forest**, a large area of land covered with many, many trees growing close together.

Almost 1/3 of all the land on Earth is covered with forests. Some are in very cold places. Others are in places that are hot all the time. The world's forests don't all look the same, but each one has many, many trees growing close together.

TAIGA

The world's largest forest runs all along the bottom edge of the tundra. This is an extremely cold forest of evergreen trees that stretches on for miles and miles, as far as the eye can see. This is the **taiga** (TYE guh). Sometimes called the **snow forest**, it covers large sections of North America, Europe and Asia.

The green areas of this map are the taiga. As in the tundra, there are only two seasons. Taiga winters are long and extremely cold, with temperatures often below zero for many months at a time.

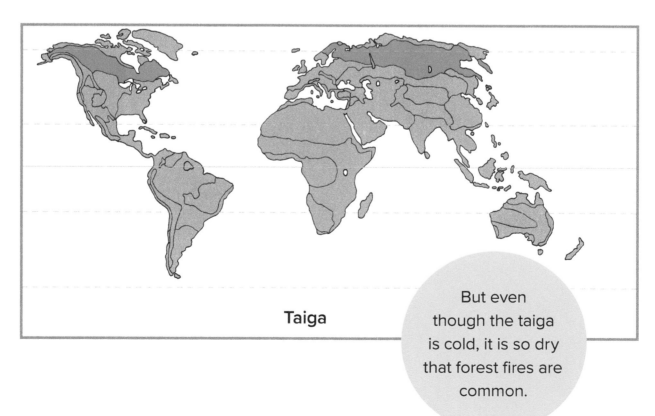

Taiga

But even though the taiga is cold, it is so dry that forest fires are common.

Summers in the taiga are short. Ferns and mosses spring up on the forest floor. As in the tundra, thousands of insects appear. Many birds migrate to the taiga in summer to build nests in the trees, hatch their young and feed on the bugs.

Animals that live in the taiga must be able to handle the extreme cold. If you were to snowshoe through the taiga, you might spot moose, black bears, wolves, foxes, bobcats, river otters, falcons and eagles.

TEMPERATE FOREST

We started our journey near the North Pole and headed south, passing first through the tundra, then the taiga.

If we were to continue heading southward, we would notice the taiga gradually changing into a different type of forest.

First you might notice that you hear many more birds.

Then you might see that instead of needles like the evergreens of the taiga, many of the trees have larger, flat leaves.

They are green in summer, turn bright colors in autumn, then fall from the trees.

Most of the trees stay bare through the winter, then grow new leaves in the spring.

You have come to the type of forests that grow in the temperate climate zone. We call them **temperate forests**.

Here in the temperate forests, winters aren't as cold as in the taiga. Summers are longer and warmer. There are four seasons instead of just two. The gradual change from winter to summer creates spring, and the gradual change from summer to winter means there is an autumn. This may be the kind of forest you are most familiar with.

The world's largest areas of temperate forest are in parts of North America, Europe and Asia.

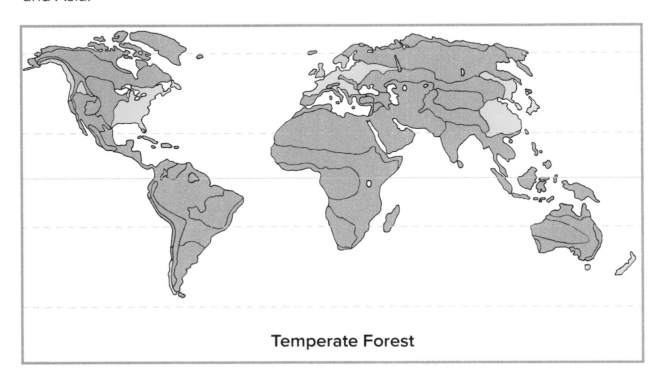
Temperate Forest

Because the climate is more comfortable, temperate forests are home to a great many more plants and animals than the tundra or taiga.

Walking around in a temperate forest, you will most often see squirrels, deer and many different kinds of birds. You might notice robins, cardinals, blue jays, woodpeckers, owls and hawks.

In the temperate forests of North America you will find rabbits, otters, beavers, raccoons, porcupines and bears. There are many small animals and insects like frogs, snakes and salamanders, flies, bees, butterflies, wasps and dragonflies.

In the temperate forests of Asia you might find deer, birds, otters, monkeys, panda bears, lynxes, leopards and Siberian tigers.

Because the weather is more inviting, many more humans live around temperate forests than in the tundra or taiga.

TROPICAL RAINFOREST

The third forest we're going to explore is even further south than the temperate forest. This is the **tropical rainforest**.

Some of the trees in a tropical rainforest grow very tall. Their leaves create a top layer called the **canopy**. A canopy is a covering. This layer gets a lot of sun. It also shades the lower layers.

Not all parts of our planet get the same amount of rain or snow. The tundra, for example, gets between 5 and 10 inches of snow per year, while the taiga gets 15 to 30 inches of snow and rain. The Sahara Desert in Africa, however, gets a total of only 3 inches of rain in an entire year!

In the tropical climate zone around the equator, an enormous amount of rain falls each year—between 6 and 30 feet!

This makes the tropical rainforest very different from the taiga and the temperate forest. It is a tall, thick, humid jungle crowded with trees and plants, and noisy with the sounds of many fascinating animals and birds.

Beneath the canopy is a layer called the **understory**. Here we find banana trees, rubber trees, avocado trees and many others.

The shadiest layer at the bottom is called the **forest floor**. So many plants and vines grow here that the rainforest is often hard to walk through.

Most of Earth's tropical rainforests are found in Central and South America, Africa, Asia and Australia.

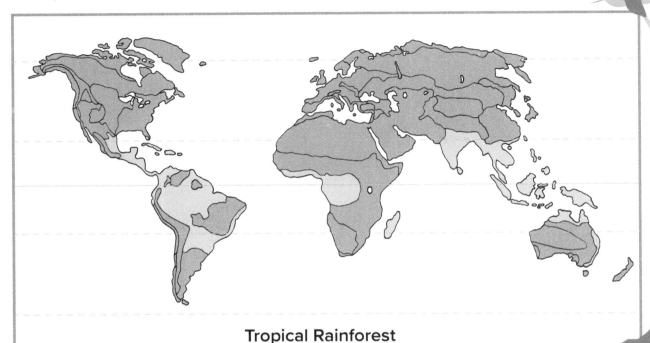

Tropical Rainforest

Half the world's animals live in its tropical rainforests! These forests are home to many of the world's most beautiful, brightly colored birds, like parrots, hummingbirds, parakeets and toucans.

Some of the world's largest snakes, such as anacondas, pythons and boa constrictors, live in the tropical rainforest, too. So do sloths, many kinds of monkeys, orangutans, mountain gorillas, tigers, jaguars, poison dart frogs and lots of different insects.

CHAPARRAL

Chaparral (shap uh RAL) is an unusual kind of forest. It looks different from any of the other forests we've talked about.

Chaparral forest is found in areas with long, hot, dry summers, and short, cool, rainy winters. The trees in chaparral forests have to go for long periods without rain, so they don't grow very tall. Grassy areas are scattered among clumps of short trees.

The land in a chaparral forest might be flat, rocky, or even mountainous. Because they are dry so much of the year, these forests have many wildfires.

Chaparral

The coasts on almost every continent have this biome. The coast of California, including Los Angeles, is a large area of chaparral.

Traveling through the California chaparral area you will find coyotes, foxes, lizards, horned toads, ladybugs, bees and birds like hummingbirds, owls and roadrunners.

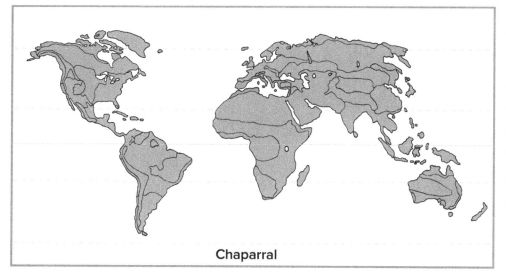

89

GRASSLAND

If you were to look at the earth from above, you would see several huge plains mostly covered with grass. We call these areas **grasslands**. They are found in places where there is enough water for grass to grow, but not enough for most trees.

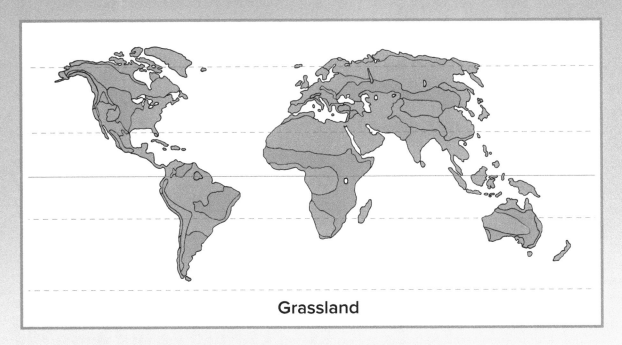

Grassland

Grassland is one of the most common biomes on Earth. It can be found on every continent except Antarctica.

Grasslands are the areas of the planet most often used for growing crops like corn and wheat, and for raising livestock like cattle.

In North America, grasslands are often called **prairies**. Here there are four seasons. On prairies you can find deer, buffalo, rabbits, prairie dogs, badgers, insects like grasshoppers and crickets, and many birds.

The large grasslands that spread across Europe and Asia are called **steppes** (steps). Wolves live there, along with bears, foxes, badgers, deer, tortoises, and small mammals like hedgehogs, wild hamsters and gerbils. Many birds live there too, including falcons, hawks and eagles.

The large grasslands of South America are called **pampas**. Living in the pampas you might find foxes, deer, skunks, guinea pigs, owls and many other birds.

Grasslands near the equator in Africa are called **savannas**. These are warm or hot all year with just two seasons, a rainy season and a dry one. Many large animals live in these grasslands—for example, elephants, rhinos, hippos, warthogs, giraffes, zebras, lions, cheetahs, hyenas and gazelle.

DESERT

Deserts are large areas of very dry land with few plants. Except for Europe, every continent has them. In fact, 1/3 of the earth's crust is desert.

Many deserts are in very warm places. Think of the Sahara Desert, for example.

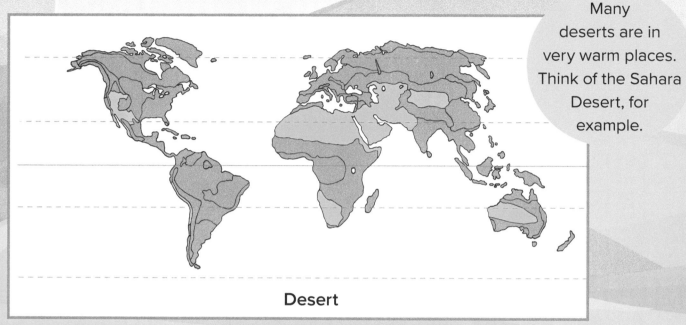

Desert

When you think of deserts, you might think they are mostly covered with sand. But sand deserts are only a small part of the world's desert areas.

Some deserts are extremely cold. Antarctica, for example, is a desert. It gets no rain at all. A small amount of snow falls each year. Antarctica is actually the driest continent!

Some deserts are much rockier, with jagged mountains, rugged boulders and many very tough plants.

The Atacama Desert, for example, is a rocky desert high in the mountains of South America. It is the driest desert on Earth with only ½ inch of rainfall each year. Some parts of the Atacama have never seen any rain at all!

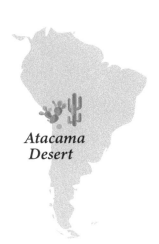

To make up for so little rain, many desert plants such as cactus and yucca store water in their stems, leaves and roots. While a desert may look brown and dry, when it does rain, the plants often bloom with beautiful flowers.

Animals that live in warm desert areas include foxes, coyotes and bobcats. There are also many kinds of snakes and lizards, geckoes, toads, and insects like beetles, grasshoppers, mosquitoes, butterflies and bees.

Many small desert animals spend their days burrowed into the earth to stay cool, then come out at night when it is no longer so hot.

CHAPTER 9: OUR AMAZING PLANET

Earth is a remarkable planet. It has a core as hot as the surface of the sun, surrounded by layers of melted rock. On top of all this is a thin, cooler crust covered with mountains and plains, rivers and oceans, forests, jungles and deserts. This crust is home to many fascinating animals and plants.

There's so much to know about our planet that some geographers spend their whole lives exploring and learning what's on it and how it all works.

Now you are more familiar with it. You know things like what it's made of, what its surface looks like, how volcanoes and earthquakes happen, what glaciers are, and why we have seasons.

We've talked about which areas of our planet are hot, and which are cold, where the main plant and animal communities are, what each is like, and much more.

What other things are there to learn and discover about our amazing planet Earth?

Well, you're a young scientist. Go find out!

WHERE IS IT?

For this activity you will need

- globe
- world atlas, or access to the internet for research

Steps

1. Use your atlas along with any other maps, books or online search you want, to find each of the following features.

2. Once you know where it is, find each one on your globe.

North America

- Appalachian Mountains
- Rocky Mountains
- Mississippi River
- The Great Plains
- Great Lakes—Lake Superior, Lake Michigan, Lake Huron, Lake Erie, Lake Ontario

South America

- Andes Mountains
- Atacama Desert
- Amazon River
- Pampas

Europe

- Ural Mountains
- Volga River
- The Alps
- Mediterranean Sea

Asia

- Himalaya Mountains
- Mt. Everest
- Yangtze River
- Gobi Desert
- The Great Steppe

Africa

- Nile River
- Sahara Desert
- Mt. Kilimanjaro
- Kalahari Desert

Australia

- The Outback
- Great Dividing Range
- Murray River
- Great Victoria Desert

Antarctica

- Southern Ocean
- Antarctic Peninsula
- Transatlantic Mountains
- Ross Sea

EXPLORE A PLACE

For this activity you will need:

- children's encyclopedia, or access to the internet for research
- your science journal

Steps

1. Choose some place on the earth that you are interested in and would like to know more about. Research the place you picked to learn more about
   - its land features
   - its climate
   - the most common biome in that area
   - the people who live there
   - their clothing and houses
   - anything else you want to know

2. Write a report about what you discover. Add illustrations if you like. Share your report with someone and try to answer any questions they have.

3. Do one of these activities with the place you researched.
   - prepare a food dish
   - learn a song or dance
   - learn and tell a story
   - make a costume
   - learn a game

4. Share the activity you did at step 3 with one or more other people.

MAP— EARTH'S BIOMES

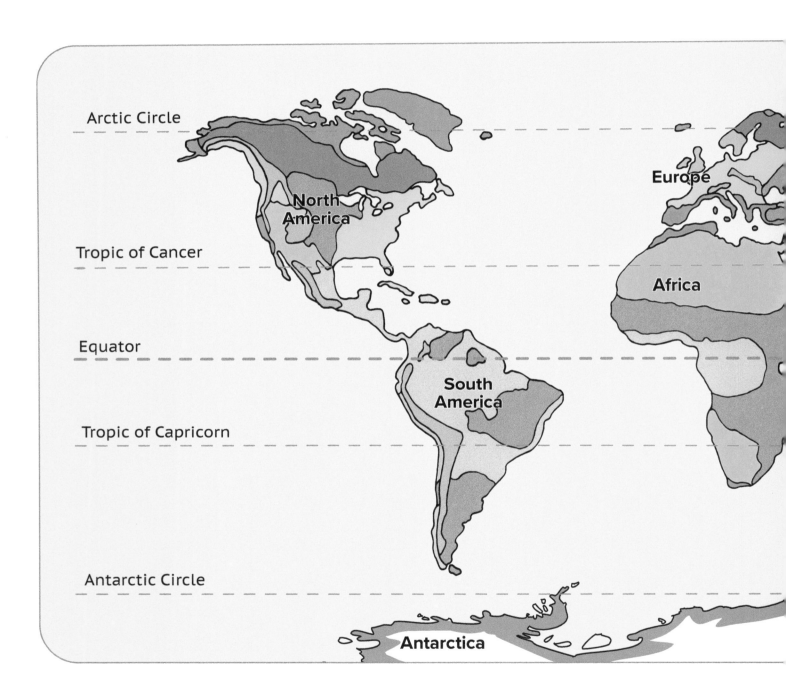

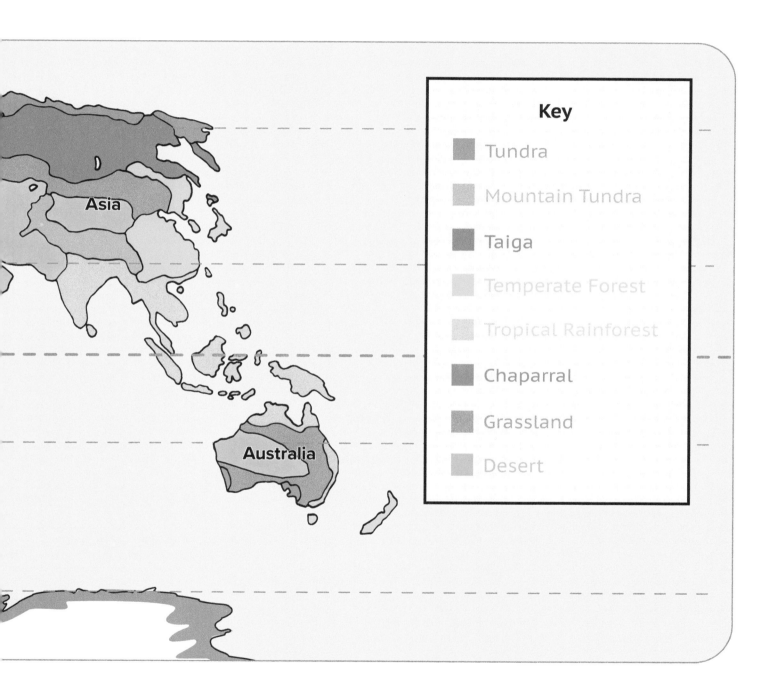